Great Steam Trains

Princess Royal and Coronation Scot

Tom Farris

Hamilton-Vale Publishing

© 2022 Tom Farris

ISBN 978-1-84285-562-1

Published by Hamilton-Vale Publishing,
New Lamorna House, Abergele LL22 7DY
www.graham-lawler.com

All rights reserved. No part of this book may be reproduced or stored in an information retrieval system without the express permission of the publishers given in writing.

The moral rights of the author have been asserted.

In the production of this book we have sourced images, some of which we understand to be copyright-free / public-domain images. This matter is then dealt with under the fair use / fair deal sections 29/30 Copyright, Designs and Patent Act 1988. In the event of a copyright claim, claimants are invited to contact the publisher, with appropriate evidence, and we will happily amend further editions.

Designed and typeset in Wales, UK
Printed in Europe.

This book is produced for educational and entertainment purposes. Readers are advised that they should not enter into any binding agreements on the basis of material in this book, without taking appropriate professional advice. Neither the author nor publisher nor any/all of the publisher's agents can be held responsible for any subsequent outcomes.

You'd think only the royal family could ride in a train with such names. It wasn't like that. Anybody could travel in one. It was all about how fast they could go. Then why had the railway picked names that had nothing to do with speed? Princesses were supposed to sit still looking posh, not race around getting sweaty!

Actually, *Princess Royal* wasn't the name of a locomotive. It was the name of a *class* of 12 of them, all pretty much alike. These big, heavy, locomotives were given dainty names like *Lady Patricia*, *Queen Maud*, and so on. Who thought to do that?

It was the *London Midland and Scottish Railway* (LMS) because it was at the time of the coronation of Prince George the sixth and his wife, Princess Elizabeth in 1937. In those days people got excited by these things probably more than they do today. At the time this railway was running trains along the *West Coast Main Line* from London to Glasgow and Edinburgh. Actually, the LMS was made up from a number of small railway firms. They had joined end-to-end to make one longer railway. It wasn't quite that simple though.

These little firms all had their own ways of doing things. They didn't always get on too well together. They argued over one thing after another, and it had started with coal.

The Start of the Arguing

In the 1830s, lots of short railways had started up all over the country. Many were for carrying coal from mines to factories. In the end there were 120 railways. One such railway was the *Midland Counties Railway* (MCR) in Derby. This is more or less in the middle of England where there was lots of coal under the ground. There isn't so much now, and we've got to leave it there anyway. Back then it was a different matter.

Not too far away is Leicester. Big factories there needed lots of coal for their steam engines. These engines drove the machines in the factories by an overhead drive shaft, pulleys and drive belts.

Overhead drive in a factory, with belts down to the machines. The drive was from a steam engine in another building, with the drive shaft coming in through a hole in a wall. (Photo frpm www.lathes.co.uk, no copyright stated. Reproduced under fair use.)

The people running the MCR wanted to carry coal from mines near them to these factories. There was already a short railway line called the *Mansfield and Pinxton* railway. This went part of the way. The idea was to make it longer. The trouble was there were coal mines nearer Leicester.

They were the only mines sending coal to the factories so they could charge what they liked for it. Having other mines sending coal as well would make the coal cheaper, they'd sell less, and they'd make a lot less money. Never mind that people were having to pay more than need be for things made in the factories. That's when the arguing started. It went on and on. Will the railway ever be built?

It was, after two years or so of arguing. Nobody got what they wanted. The new railway used only part of the *Mansfield and Pinxton* railway. Using all of it had been the reason for the railway in the first place. So did this make the mines nearer Leicester happier?

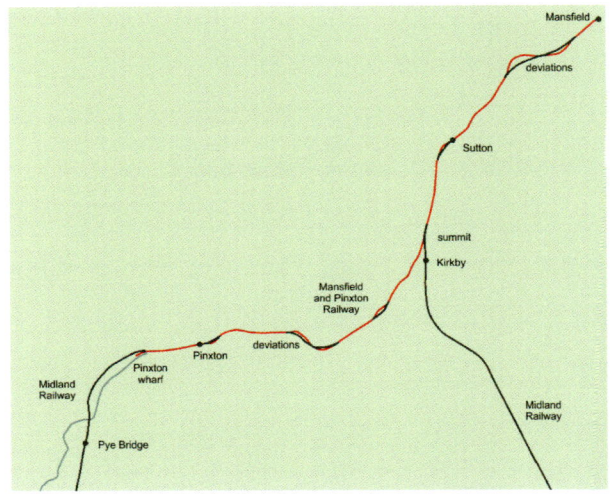

Map showing the Macklesfield and Pinxton railway (in orange) and the Midland Railway lines (in blue). This was later on in 1851 when the MCR was part of the Midland Railway, and a new line was built to Pinxton. (Image Afterbrunel, Creative Commons Attribution-Share Alike 4.0 International license.)

It didn't. They were just as unhappy because the railway caused the price of coal to drop, so they made less money. There was one thing though that the MCR wasn't to know at the time

It was one of the little railways that led much later on to something a lot bigger. This was the *London, Midland and Scottish Railway* (LMS), the *Princess Royal Class* and the *Coronation Class* locomotives, as well as the *Coronation Scot* train service. These were a long way off though. In the meantime, the changes that would lead to this had started to happen.

Soon after that unhappy railway was built, the MCR joined up with other railways. The whole lot became the *Midland Railway*. This was good for them because they ended up being the biggest railway company in the country, running trains from Derby in many directions over England. They wanted to run trains to Scotland too, but there were two other railways that wanted to as well. The one that worried the Midland Railway more was the *London and North Western Railway* (LNWR).

Backtrack: the Cotton Railway

The LNWR railway had started before the MCR as the *Liverpool & Manchester Railway* (L&MR) back in 1830. It was the first passenger railway ever to use steam locomotives instead of horses. It was built to save rich people money.

Cotton coming in by ship to Liverpool docks was being carried by canal to Manchester where it was made into clothes in big, noisy, factories. Working in these factories was risky and hard. The bosses treated the workers so badly that many went deaf from the noise, died young or got killed by the machines. Children were made to climb over machines while they were running to keep them clean. Many got badly hurt or killed too. Why did the bosses treat their workers so badly?

Showing people what it was like for the workers, The Manchester Museum of Science and Industry has a big cotton-spinning workshop. The machines are shown actually spinning cotton.

These spinning machines look harmless, but they were far from it. Every few seconds, the bottom part of a machine shot out to the end of the frames that stick out to the right of the picture, then shot back again. While it was doing this, a child was made to crawl underneath to pick up anything that had fallen on the floor. If it wasn't quick enough, the machine would hit it without stopping and injure it badly. Those children must had terrible nightmares. (Photo T. Farris.)

Greed and Misery

The bosses treated workers so badly because there were no laws stopping them doing it. In fact, the cotton-mill owners and other powerful men made the laws. The workers didn't get a chance. While the working people suffered, the cotton-mill bosses made as much money as they could for themselves. They were always trying to find ways to make even more.

One way was to save money was in the carrying of cotton one way and clothes back again between Liverpool and Manchester. This had been done by canal boats, but the canal bosses were charging a lot of money for doing it. In the end it didn't do them any good.

This is because it got the cotton-mill bosses thinking about a railway. If the canal people hadn't been so greedy, the *Liverpool & Manchester Railway* may never have been built. Well, it was built. The owners were proud of the railway too, but pride can lead to a big fall.

The opening day of Britain's very first steam passenger railway was to be a grand affair, but it didn't quite turn out as planned.

The Grand Opening of the World's First Steam Passenger Railway and How It Kept Going Wrong

A drawing of the L&MR train. (Image: Public Domain.)

A painting of the opening of the Liverpool & Manchester Railway in 1830. (Image: Public Domain.)

The First Crisis: a Crash

Eight trains were to run together from Liverpool to Manchester. Seven were to run in a procession led by *Rocket* on one track. One other train carried the famous Sir Arthur Wellesley, 1st Duke of

Wellington. He was the prime minister, no less. His train set off first and ran on another track on its own.

Sir Arthur Wellesley, 1st Duke of Wellington. He was Prime Minister at the time, but not much longer. *(Photo in the Public Domain.)*

Wellington had many guests with him in a special carriage with a roof and doors.

This train was pulled by the *Northumbrian*, driven by the great railway engineer who had designed it. This was George Stephenson. The locomotive looked like the *Rocket* but that's as far as it went. It was the most advanced locomotive yet.

The 'Northumbrian' locomotive. This pulled the Prime Minister's train for the opening of the Liverpool and Manchester Railway. *(Image Public Domain.)*

That first thing to go wrong was after 21km (13 miles). This had nothing to do with the *Northumbrian*. It was one of the locomotives following the *Rocket*. It came off the rails and suddenly stopped. The driver of the locomotive following it was taken by surprise. He couldn't stop his train in time. It ran into the back of the stopped train. Since nobody was hurt and there was no damage, the locomotive was levered back onto the rails and the trains carried on. It was the world's first passenger-train crash. It wasn't a good start to the day.

Next, Death on the Tracks

The Prime Minister's train stopped as planned at Parkside station near Newton-le-Willows (in the North-West of England) to take on coal and water. While there, the trains coming up on the other track were to pass proudly by. The first one was pulled by the *Rocket*, driven by Joseph Locke.

At the time Locke was George Stephenson's assistant, but Stephenson hated him. This was because Stephenson had made a mess of planning the route for the Liverpool and Manchester Railway. Locke wrote a report saying it was a mess. He came up with a better route. Stephenson was furious. Locke's route was used. Then Stephenson had to drive the *Northumbrian* on it, which must have made him feel sick.

Before the *Rocket* got to the meeting point, about 50 people had got out of the carriages onto the tracks to stretch their legs. One of them was William Huskinson. He was the member of parliament for Liverpool. He soon wished he hadn't got out. It landed him in terrible trouble.

Along with many others he strolled back and to along the other track. Then someone called out, "Train coming!" Some people got back in the carriages. Others went over to the other side of the track. Huskinson panicked. He first went one way, then the other. At the last moment he decided to climb back into the carriage. He got hold of the door to help pull himself up before getting in, but it swung open.

Rocket hit the open door. Huskinson was left hanging with one hand onto the handle. The door was flung round, throwing him to the ground between the tracks but one of his legs fell over a rail in front of *Rocket*. Seeing what was about to happen, Locke had thrown *Rocket* into reverse, but this took ten seconds to work.

It couldn't be helped. It happened so fast. *Rocket* ran over Huskinson's leg, crushing and cutting it.

Huskinson's was already not well. His doctor had warned him not to make this trip. Lying on the ground, he is supposed to have said, "I have met my death—God forgive me!" Even though he must have been in terrible pain, he kept living. The locomotive and first carriage were taken off the Prime Minister's train. Huskinson was put on a door that had been torn off and rushed in the first carriage to the vicarage in Eccles, near Manchester, where a doctor saw him.

The only thing he could do was give Huskinson dose-after-dose of *laudenam*. This is a strong painkiller. That evening Huskinson died. He became famous all around the world for being the first person to die in a railway accident. Poor man, what a way to become famous! (Many men must have already died over the years on railways for coal mines). That was bad enough, but more trouble was to come that day.

Stephenson's famous Rocket in the Manchester Museum of Science and Industry. (Photo T. Farris.)

Rocket was to continue being used, though the memory of that accident would haunt it for ever more. It now sits on display in the Manchester Museum of Science and Industry. It seems to still have the wheels that had run over William Huskinson.

Carry On!

The happiness that everyone had been feeling on that opening day crashed into horror and sadness. The Prime Minister wanted to give up the trip and go back to Liverpool. Then news arrived by men on horses that there was a large crowd waiting for him in Manchester. He wasn't told why.

If he didn't turn up, there would be a riot. He decided it would be better to carry on. The only thing was, the locomotive that had been pulling the train was many kilometres away having taken Huskinson to be treated. All the other locomotives by the prime minister's train were on the other track. None of them could be moved to the track that the prime minister's train was on. There was only one thing for it.

Someone found a long chain. One end was tied to the prime minister's carriage on one track, the other to a locomotive on the other track. Another locomotive was brought up to help. So the journey carried on with the Prime Minister's train being pulled by a long chain stretching from one track to another alongside the carriages of the towing train. Arriving in Manchester like that would have had people making fun of them. Luckily for the prime minister, it didn't happen that way.

The Riot

Along the way they met the train's own locomotive, *Northumbrian*, coming back with the carriage, still driven by George Stephenson. He was bragging that he'd broken the speed record, getting Huskinsor to Eccles. The locomotive and the carriage with it were quickly coupled to the train and they set off as fast as they could to Manchester to meet the crowd. How pleased would they be to see their Prime Minister? Maybe that's what he was thinking. He got a surprise.

Some people cheered. Most booed and threw vegetables at his carriage. While his guests had got out to have a meal, he was too frightened to do the same. He stayed inside and had his meal brought to him. (He was hated so much he wasn't prime minister much longer.) Soon they had to go back to Liverpool, but that was the next problem.

With all the trains there, there wasn't space on another track to run *Northumbrian* around to the other end of the carriages to pull the train back to Liverpool. The people crowding over the tracks got in the way too. In the end, the prime minister's train managed to leave. So did the other carriages. It was all such a mess that they were towed by ropes from three locomotives. People must have laughed and jeered. Even that wasn't the end of the railway's troubles.

What More Trouble?

The trains should have got back to Liverpool at 4 o'clock in the afternoon. They were nowhere near. At 7 o'clock they were crawling along only halfway. It started to rain. Some of the carriages were open wagons, so the people in them got cold and wet. So did the locomotive crews. Then it got dark. There were no lights on the trains so a driver set fire to a rope and walked in front to give some light.

Crowds along the way had been waiting a long time. Many of them were drunk. Some on bridges threw things down at people on the train. Somebody had put a wheelbarrow on the line. Then the train came across a group of people in uniform stumbling along. At first the crew and passengers thought they were the army, It wasn't. It was the members of the band that had been playing until the accident happened.

Because their carriage had been used to carry Huskinson to Eccles, they had to walk home, now in the dark and in the rain. They were still walking. The people on the train didn't have it much easier either.

A modern brass band during COVID lockdown in 2020, hence the blackboard in place of the band's leader.
(Photo from the Big Issue.)

The last part of the journey was uphill. The three locomotives pulling the train of 24 carriages couldn't pull them up the hill. To make the load less all those important men had to get out and walk for a mile in the rain in the dark while the locomotives struggled to pull the empty train up that hill. Very, very late, the trains arrived in Liverpool. A grand celebration was laid on. Nobody was in the mood. They just went quietly home.

Not Quite As Planned

This was how the grand opening of the world's first steam-powered passenger railway worked out.

North to Scotland

The Midland Railway wanted to go to Scotland. The only way there for them was on the west side of the country through Carlisle. This town is on the English side of the border with Scotland. At the time, there was just one line running to Carlisle on the English side. This was owned by the LNWR. They wouldn't let the Midland Railway use it. What could the Midland Railway do?

They built their own line to Carlisle. Some of the land on the way there is very hilly. The LNWR line had taken a long way round to keep off the hills. The Midland Railway decided to go a much shorter way over them. This became the *Settle and Carlisle Railway*. It was not an easy railway to build.

First it climbed up to higher ground. Then it crossed the valley of the River Ribble. It's at a high, lonely, windy, rainy, place on a bleak moor near the Lancashire-Yorkshire border. The only way to cross it was by building a long viaduct. This is the famous *Ribblehead Viaduct* with its 24 arches 32 metres (104 feet) tall.

It was tough on the workmen building it, often in horrible weather. People sitting in comfy trains crawling over the viaduct, and people taking photos of it, don't for a moment think about how tough the work was building it, and how many men died.

South to England

North of the border, after much arguing among the Scottish railway companies, the *Caledonian Railway* was allowed to build their railway into Carlisle. They then wanted the *Glasgow, Paisley and Greenock Railway* (GP&GR) to join them. This went ahead without the Caledonian having to pay any money to the owners of the GP&GR. Instead, these owners would get money each month once the railway was earning it The problem in the beginning was there wasn't enough money coming in to pay these owners. They got angry. At the same time, the Caledonian Railway people noticed that some of the owners of the GP&GR were cheating the firm of money. It was so bad, these owners had to leave the GP&GR.

The Caledonian Railway started running trains from Glasgow and Edinburgh to Carlisle. On the English side, trains were running from London to Carlisle. For the first time, people could travel between London and Glasgow or Edinburgh. This was in 1848.

As the years went by more railway lines were built going into Carlisle from the north and the south. By 1912, there were nine of them.

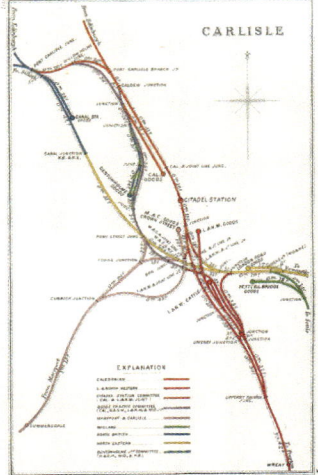

The nine railways running into Carlisle in 1912.
(Image Public Domain.)

Why Glasgow?

Why run trains from the capital city of England to just a city in Scotland? After all, the capital city of Scotland is Edinburgh, not Glasgow. It's because Glasgow had something that Edinburgh never had.

Looking at the city now you would find it hard to believe why it's called *Glasgow*. The name comes from old words that roughly mean 'green enclosed place' (*glas gau*). This is because when people first found it, it was a beautiful, green, place on the edge of the River Clyde. That changed when the *Tobacco Lords*, as they were called, moved in. They built harbours and deepened the river so that their ships could sail all the way there to unload their cargoes of tobacco. Soon the whole country was puffing away on cigarettes, cigars and pipes. The Tobacco Lords became richer and richer as the people became sicker and sicker from breathing in the smoke from burning the Tobacco Lords' poisonous leaves.

In time, ships began to be built in Glasgow, then railway locomotives. It went on. Chemicals, industrial machinery, bridge building, explosives, coal and oil industries, textiles, clothes, carpets, leather, furniture, pottery, food, drink, cigarettes of course, printing and publishing. What else could a city need?

A railway, of course. Glasgow had

become a major city. This is why the railway companies were working to connect Glasgow with London. Even without the problem of Carlisle, it wasn't easy.

Problem hills

Now that the LMS was using the Midland Railways' hilly route, the Scottish and English railways both had the same problem. This was big, steep, hills. On the English side is *Shap Fell* near the Lake District on the *Carlisle and Settle Railway*. This is two thirds of the way between the towns of Kendal and Penrith in the North-West of England. The climb up is rather steep for trains heading north. The early locomotives could only pull short trains up them, and then they needed help. This was with a *banking engine* that would push the train from behind.

On the Scottish side is *Beattock Summit*. It's about 80 km (50 miles) south of Glasgow. As with *Shap Fell*, trains travelling north needed help from a banking engine. These were kept in an engine shed at Beattock village.

On these climbs the locomotives used steam very quickly. The firemen would be shovelling coal as fast as they could as well as keeping the water in the boilers topped up. If a fireman didn't work so hard, the locomotives would have more trouble climbing the hills. A steam locomotive was only as good as its fireman.

Slow and Fiddly

The railway line was built when many express locomotives of the day had either one big driving wheel each side, called a '2-2-2 or 4-2-2 single wheeler', or two smaller driving wheels in a 4-4-0 layout. Which was better?

A Caledonian Railways single-wheeler locomotive from around 1880. *(Photo Adrian Covic via Pinterest.)*

The bigger driving wheels made the locomotive go faster, but they weren't so good on the hills. The other locomotives were better at climbing hills but weren't so fast. Which type of locomotive should the railway companies use?

They used both. They could do this because no locomotive could pull a train the whole way without running out of coal and water. Also the driver and fireman couldn't work that long either. So locomotives with their crews were changed on the way. In those early days the railway companies used the faster single-wheelers on the flat land, and changed to 4-4-0 locomotives for the hills. In time, they found the single-wheelers they had ran roughly on the track and came off the rails easily, so they rebuilt some as 4-4-0 locomotives. Changing locomotives to another type happened quite a lot.

A beautiful Midland Railway 4-4-0 locomotive from 1893. (Photo in the Public Domain.)

These locomotives were owned by the little railway companies that made up the LNWR. They carried on for around 70 years. The companies saw no reason to change. It wasn't just the locomotives they didn't want to change, it was the way they worked too. Since the different LNWR companies worked in different ways, they couldn't stop arguing. It meant passengers suffered as a result.

In 1914, the government noticed. With the First World War about to start, the government was probably worried that thousands of soldiers could be stranded waiting on station platforms for trains to Liverpool docks on their way to fight in the war.

The government forced these railway firms to become just one, carrying on with the name LNWR. What it didn't do is stop the arguing though.

The people from the little firms still ran their parts of the LNWR their own way. In doing so, they didn't care how much trouble they gave their passengers. Something more had to be done.

It took another nine years before anything was done. In 1923, the government tidied up the railways some more. With the LNWR, Midland Railway and more English and Scottish railways brought together, a new railway company was formed, This was the *London, Midland and Scottish Railway* (LMS). Time for new thinking?

The people who used to run the small railways that became part of the LMS still thought having lots of small locomotives was still the right way. Meanwhile, others could see that the LMS needed some big locomotives to pull proper express trains to and from Scotland. After all, that's what the east coast railway was doing by then with Nigel Gresley's *Pacific Class* locomotives. The famous *Flying Scotsman* was one of them. The LMS locomotives were looking out of date, even if some of them were rather beautiful.

A more powerful locomotive would pull trains up Shap Fell and Beattock Sumnit without messing about with banking engines. In all, trains would travel faster. Journey times would be shorter. People would be less tired from travelling. They'd be proud to be travelling on a smart, modern, train.

Model of a Midland Railway '6 Class' 0-4-4T tank engine. The real locomotive was on loan to the Midland and Great Northern Railway (M&GN) *(Sales photo by, 7mm oco.co.uk, copyright not stated.)*

The LMS train drivers battled on, squeezing what they could out of their low power locomotives. To get more speed, a second locomotive would be attached at the front. This is called *double heading*.

Railway lovers get excited to see this, but it shows how little power the locomotives had for what they were needed to do.

Wake-up Call

Nigel Gresley's *Pacific Class* locomotives were grabbing all the glory for trains running to Scotland, and most of the passengers too. At last, a wake-up call was waiting for the LMS.

It still took them eight years to see that it really was time to have locomotives to rival the LNER's *Pacific Class* locomotives. They just needed someone to design one for them. Someone like a Chief Mechanical Engineer from another railway, but these don't usually just hand in their notice and go. What was the answer?

It came from the very top of the LMS: Sir Josiah Stamp, no less. He was Chairman (big boss) of the LMS and, among many other things, a boss at the Bank of England. He spoke to William Stanier who was Works Manager for the Great Western Railway (GWR). Stanier was working for their Chief Mechanical Engineer, George Churchward. He had designed the very best locomotives in Britain at the time. Would Stanier like to leave the GWR to become Chief Mechanical Engineer of the LMS and design new locomotives for them?

You bet! He started work on the 1st January 1932, bringing with him great ideas he'd learnt from Churchward and engineers before him. Stanier built teams of engineers, and office support staff, at the Crewe Works. They got down to work on the first locomotive to show the LNER a thing or two.

Who was William Stanier?

William Stanier had followed his father into working on the Great Western Railway in their vast Locomotive Works in Swindon, 130 km (81 miles) west of London. His first job was as an office boy, running errands, making tea, and so on. After

(Photo used under Creative Commons Attribution-ShareAlike License)

a while of tripping over desk legs and spilling tea, he spent five years as an apprentice (learner) in the workshops. By this time he knew quite a lot about how steam locomotives were put together. He was then trusted to start designing bits of them in the Drawing Office. There his job was *Draughtsman*. Maybe he wasn't quite ready to be designing but must have shown he was very thorough in his work because after just three years he was given a different kind of job.

It was *Inspector of Materials*. This was to make sure the steel and other materials that the workshops bought were up to scratch. Another three years later, he was given another leg up as *Assistant to the Divisional Locomotive Superintendent* in London. His life seemed to go in steps of three years because in 1912 he was sent back to Swindon where he became *Assistant Works Manager*. This job lasted eight years until he was made *Works Manager* working directly under the 'grand master' of locomotive design, George Churchward. No wonder Sir Josiah Stamp wanted Stanier for the LMS.

Sorting out the LMS

The first thing Stanier did was kill the idea of small locomotives for pulling express trains. With Josiah Stamp's backing he went all out to design a *class* of fast, powerful, express locomotives. They would be the pride of the LMS. Where to start though?

LNWR's beefy Dreadnought Class locomotive with the dainty name of 'Marchioness of Stafford'. Dreadnought locomotives were built between 1884 and 1888.
(Photo in the Public Domain.)

There was no point starting from scratch when he knew about the GWR's express locomotives. The latest one of these was the 4-6-0 *King Class*. He took that as his starting point. These were the GWR's and Britain's most powerful and biggest express locomotives at the time.

Could he do better?

He did. The next year with his guidance, his engineers had designed and built a new locomotive. Not only that, they had squeezed out more power than the *King Class* had. Stanier had told them to fit a bigger firebox and make other changes from the *King Class*. This firebox was heavier than the one in the King Class, so they had to add extra small wheels to hold the rear of the locomotive up better. That meant a 4-6-2 wheel layout like the LNER's *Pacific Class*. They built just two to begin with. These locomotives were ready for the LMS's express trains.

This was good because the LMS wanted these locomotives to pull long trains on the new so-called non-stop *Coronation Scot* train service between London and Glasgow. The trains did stop though. This was just once at Carlisle for locomotives to be changed. This was the only station on the way that passengers could get on and off.

> The numbers 4-6-2 and so on are called the 'Whyte notation'. The first number is the number of small wheels before the big driving wheels, the middle one is the number of driving wheels, and the last is the number of wheels after the driving wheels. Small ones at the front take the weight of the front of the locomotive and help to steer it on the tracks. Small ones at the back take the weight of the back of the locomotive.

The LMS was so proud that they thought up that name, *Princess Royal*, for the class. They called the first locomotive *The Princess Royal* probably to make sure people didn't forget the name. The second one was *Princess Elizabeth*. Then there were another ten with names like these. None of this royal talk did anything for the engineers in the railway workshops. They simply called them *Lizzies* after the *Princess Elizabeth* locomotive.

In 1936, the LMS started their non-stop train service to Scotland and back, giving it the name *Coronation Scot*. Neither this name or *Princess Royal* stood for speed (as the rival LNER's *Flying Scotsman* had done). Maybe the LMS were aiming at the snobs. "Oh, we travelled on the *Coronation Scot* and it was pulled by *The Princess Royal* don't you know!" Whatever it was, it worked. The train service made the LMS a lot of money.

Problem Hills—No Problem

On the way are those two problem hills: Shap Fell, and Beattock summit. Instead of struggling up them with a banking engine pushing from behind, the Princess Royals whizzed up them under their own steam. Problem solved, but Stanier had wanted to try a better idea.

Really Non-stop This Time

In 1928, the rival LNER started their non-stop service from London to Edinburgh. They were so proud of this, and made sure everyone knew. All well and good, except the LMS had beaten them to it. Four days beforehand, they made their own proper non-stop run between the two cities. How did they do it?

First of all, thanks to Stanier's ideas, the *Princess Royal Class* locomotives could run the whole way with just the coal in the tender. Water troughs could fill the water tank in a special tender. This tender had something else that made it special too.

It was called a *corridor tender*. A corridor (tunnel) allowed crews to go between the footplate and first carriage where they rested until they were needed for driving on the tracks they knew. (For more on the corridor tender and water troughs, see Tom Farris's book *Flying Scotsman*.) This didn't solve a long-standing problem though.

Hammer Blows

When a locomotive is going fast, the pistons go back-and-to in the steam cylinders like hammers. These hammer blows travel through the wheels onto the rails and down into the track bed underneath. This is bad for the track bed, even more so for bridges. Trains had to go slowly and gently over them, otherwise the hammer blows could make the bridge break up and fall down. Was there a way to stop the hammer blows? That was a question Stanier and many other locomotive engineers were asking.

Nobody in Britain had the answer but a firm in Sweden did. This was called the *Ljungström Turbine Company*. The clue is in the name *Turbine*. Instead of pistons bashing back-and-to in cylinders there's a *turbine* that does no more than spin smoothly round very fast. A turbine is like a fan with lots of blades in several rows. In this case there were 18 of them. To make it go round, high-pressure steam is squirted from nozzles around the turbine.

Stanier told his engineers that the third *Princess Royal* locomotive was to be built using a turbine in the tender instead of pistons in cylinders. The turbine was to drive six small wheels under the tender through gears. The wheels under the long boiler of the locomotive would be just small ones too for carrying the weight. Instead of the locomotive pulling the tender, the tender was pushing the locomotive.

When built, Stanier was eager to find out how good it was. He was pleased to find it used less coal than a normal locomotive. There was a downside to this though.

Turbine powered locomotive.
(Photo Tekniska museet Stockholm, Public Domain.)

The locomotive needed two turbines, one one for going forwards, and one for going backward. There was only enough space for a small one for going backwards, so the locomotive had too little power for pulling trains with the locomotive going that way. This is called *tender first running*. Crews didn't like this anyway.

This was because coal dust and rain could pour into the cab. In some cases, a storm sheet was used to cover the gap between the cab roof and the tender to keep rain out. Just as bad, it was hard for the driver to see past the tender. Even so, the LMS used the turbine locomotive for around 14 years until the turbine stopped working. It would have cost too much to fix, so t was stripped out and the locomotive rebuilt with cylinders. It was then named *Princess Anne*. Sadly, it was to suffer a terrible fate soon after.

In just two months of being back on the rails, the locomotive was caught up in a horrific rail crash with two other trains. It wasn't the fault of the *Princess Anne*. This was in October 1952 at Harrow and Wealdstone station in North London. Some people died and others were badly injured. The locomotive was so badly damaged it couldn't be repaired.

'Princess Anne' on its side after the terrible crash in October 1952. (Photo Public Domain.)

This was because of the mistake of a driver of another train. He and his fireman were killed too. Nobody knows why the driver made that terrible mistake. The crash could have been avoided if the railways hadn't been behind the times with safety equipment.

An Automatic Warning System box between the rails, (Photo Oxyman, licensed under Creative Commons Attribution-Share Alike 3.0 Unported license.)

Soon after, the government made railway companies fit *Automatic Warning System* boxes by signals. These were yellow boxes fixed between the rails. They set off a warning in the cab if a driver carried on past a stop signal. There was a lot of argument against them.

One was that not many people had been killed in railway accidents over the years, so why bother? Another argument was that drivers knew what they're doing. They didn't need gadgets to tell them what to do. If they had been fitted at the signals for Harrow and Wealdsone station, that accident may not have happened.

In 1936, well before that awful crash, the LMS had decided they needed more *Princess Royal* locomotives for the *Coronation Scot* train service.

Bigger and Better, But...

Stanier, planned to build five more, but the Chief Draughtsman, Tom Coleman, said it would be better to design a new class of locomotive that was more powerful, more reliable and easier to work on. Stanier went along with the idea, but he had to go to India before work started. He handed the design work to Coleman, who did a fantastic job.

As well as making some parts inside work better, the only way to make a steam locomotive more powerful is to make it bigger. It can only be made so big because it has to pass under bridges, pass other trains and run alongside station platforms without hitting any of them. This size is called the *loading gauge*. Different railways have their own loading gauges.

Some might be narrower, others lower, and others smaller all round. For instance, London Underground *Tube* trains just fit inside their tunnels with a small, round, loading gauge.

(Photo SPSmiler, Public Domain.)

Coleman made the firebox, boiler, cylinders and wheels bigger, so much so that the locomotive was so big it only just fitted in the loading gauge for the lines it was to run on. He also did something to help the fireman.

This was a steam powered *coal pusher*

that moved the coal in the tender to the front. This saved firemen walking back-and-to further into the tender as coal was being used. Then a blow was struck.

All was going well. The Drawing Office had nearly finished designing the locomotive. The workshop was getting ready to build the first one. A name had been decided for the class. Not giving up on royal titles, they came up with *Coronation Class*. Then the Marketing Department said they wanted it *streamlined*. This was because Nigel Gresley's engineers on the rival LNER had come up with the streamlined A4-Pacific. The famous locomotive *Mallard* was one of them.

LNER's Mallard (Photo PTG Dudva Creative Commons Attribution-Share Alike 3.0 Unported.)

"Streamlined? There's no room left for adding streamlining! Also it could make the locomotives too heavy for the track." The Marketing Department said it had to be done, no argument. "You'll have to find a way to do it." Designers hate somebody coming along wanting big changes when they have got everything just right. It meant having meetings to work out what parts can be changed and by how much. It also meant getting slide rules out of desk drawers to go through some of the maths again. After that, new drawings had to be made for the workshop to follow. If a locomotive has already been built, it had to be taken apart, the new parts fitted and put back together again. It's amazing what you can do when pushed though.

They found a way to wrap streamlining panels tightly around the locomotive without going out-of-gauge. This wasn't the end of the problems though. The streamlining weighed around 5.5 tonnes (5 tons). The designers had to find ways to make the locomotive itself that much lighter. By taking a bit off here and a bit off there, they managed it. The new locomotive was now streamlined. The Marketing Department loved it. The Engineering Workshop hated it. The designers probably did too.

The streamlining panels made looking after the locomotive much harder to do. Never mind that, it made people notice. The LMS was now at the leading edge of locomotive design. They had Britain's most powerful steam locomotive, and it was streamlined, so it must be fast too.

It was, but it wasn't the fastest. The LNER's *Mallard* held that title. There was a new problem though. This was a problem that a lot of steam locomotives had.

This was smoke from the chimney blowing down the sides of the locomotive and blocking the driver's view of the track ahead. The LNER's streamlined A4-Pacifics didn't do this thanks to a careless thumbprint made on a model when the engineers were testing it in a wind tunnel. No such luck with the LMS's new streamlined locomotive. Drivers had to suffer having to peer through smoke blowing down the sides of the locomotive. This was too risky. Something had to be done about it.

With the engineering workshop as well as the drivers now hating the streamlining, there was only one thing to do. The workshop took it off all ten that had it fitted. All new locomotives were built without it. The smoke problem was solved the usual way with *smoke deflectors* at the front of the locomotives.

'Duchess_of_Hamilton, Coronation class' locomotive with slight streamlining and with smoke deflectors
(Photo Voice of Clam, Public Domain.)

Streamlining only helped above 145 km/h (90 mph) anyway. The trains rarely went that fast. That was one failed idea from the Marketing Department. Then came another failed idea, this time from the Engineering Department. This one put an end to the *Coronation Class* altogether.

It was electric trains. Not the trains themselves but the overhead wires that were put up to power them. The *Coronation Class* locomotives were so big their tops were too close to the overhead wires for safety. Worse ws to come.

The track from London to Crewe was the first part to be made electric. Since this was at the start and end of the *Coronation Scot's* journey, it meant these locomotives could no longer pull the *Coronation Scot* out of, and back into, London. That was it.

They were sent to the scrap yard to be cut up. Three were saved though.

Overhead wires that killed the Coronation Class locomotives. (Photo by Bhaskaranaidu, licensed under the Creative Commons Attribution-Share Alike 3.0 Unported license.)

Afterwards, the *Coronation Scot* rail service used diesel locomotives. Then that out-of-date name was scrapped too. The *Coronation* era had come to an end.

For more fascinating *Great Steam Train* stories from Tom Farris, see…

**Britain's First Railways
Flying Scotsman
Trains Racing North
Mallard and the A4-Pacifics
The Grand Dream of Broad Gauge**

Hamilton-Vale books are available via all good bookshops or online from Gwales.com, Amazon.co.uk and other website sellers.

***Be the first* to know about our new titles. Join our mailing list. Send an email with '*JOIN*' in the subject to info@graham-lawler.com**